Talking Planets
COMIC BOOK

The Night Problem

Written By **Iada Scott**

Talking Planets

Typesetting, Book Layout, Editing and Cover Design by Enger Lanier Taylor for In Due Season Publishing

Published By: In Due Season Publishing
               Huntsville, Alabama
               indueseasonpublishing@gmail.com
               www.indueseasonpublishing.com

ISBN-13: 978-0999238776
ISBN-10: 0999238779

Original concept and drawings for all illustrations
by Jada Scott

Presented To

_____

THANK YOU

I would like to thank mom, dad, The Adventure Science Center, and The U.S. Space and Rocket Center in Huntsville, Alabama, for the inspiration in writing my first book. You have helped me accomplish a difficult task.

Future Engineer

My name is Jada Scott and I am 9 years old. I live in Tennessee where I attend Homer Pittard Campus School. I love to study science and planets. I was inspired to write about planets and the solar system from my visit to the U.S. Space and Rocket Center in Huntsville, Alabama.

I aspire to work for NASA as a Rocket Ship Designer/Engineer when I grow up. In addition to my love for science, I enjoy art and creating solar system models.

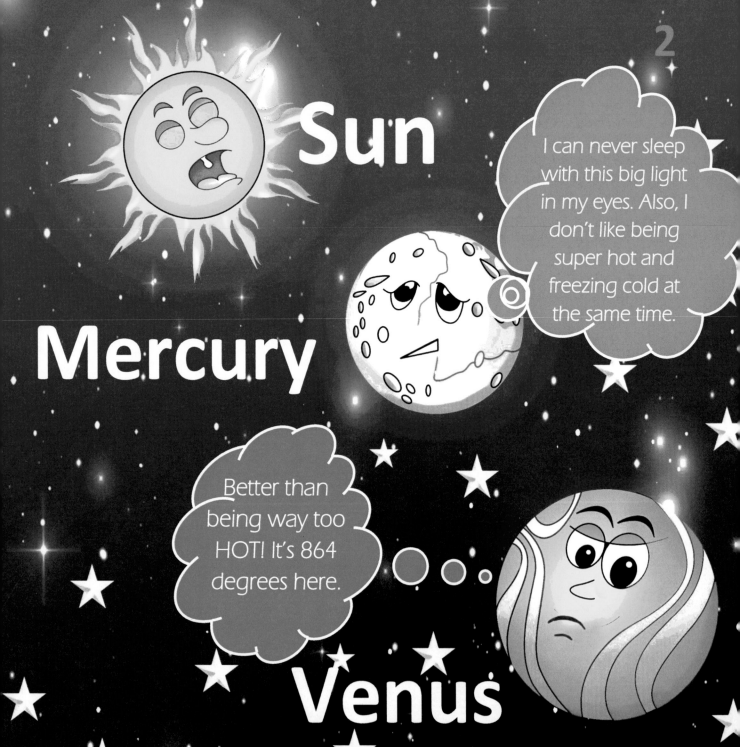

You know I'm far away from the planets, but sometimes I just want to suck them up so they will be quiet.

Black Hole

Now you know what the planets do at night. People can't really believe how it can be that loud in outer space, and why we can't hear a thing. Maybe you can solve that mystery.

Til Next Time!

Take a ride on Jada's space bus! Jada wants you to continue to learn about our solar system and the planets.

This is one of her art creations from school. She used metal, markers, and paper to create the bus. Now you can create it too and go on a journey through space.